CONTENTS

For thousands of years stargazers knew of only five planets besides the Earth. They could see these bodies with the naked eye, looking like bright stars wandering across the night sky. It was not until 1781 that another planet was spotted, which was named Uranus. The discovery of Uranus set astronomers looking for other new planets, and two more were eventually found – Neptune in 1846 and Pluto in 1930.

Uranus and Neptune are large planets, each about four times bigger across than the Earth. And they are quite different from the Earth. Like the giant planets Jupiter and Saturn, they are made up mainly of gas and liquid. Also like those giants, they have rings and many moons circling round them.

Pluto is a vastly different world from Uranus and Neptune and indeed from all the other planets in our Solar System. It is a tiny ball of rock and ice, smaller even than our own Moon. Astronomers are not sure whether it should be called a planet at all! It is so far away that it takes nearly 250 years to circle once around the Sun.

At present, Pluto is the most distant planet we know. But some astronomers are convinced that other planets exist far beyond Pluto.

FAR, FAR AWAY

Little was known about the far planets before the Space Age began.

Saturn is the most distant planet that we can see easily with the naked eye. Uranus, the nearest of the far planets, is much smaller than Saturn and very much farther away. But it can, just, be seen with the naked eye under ideal viewing conditions. It looks like a very faint star.

A telescope shows Uranus as a round disc, with a bluish-green colour. But it does not show any markings, such as the bands that can be seen on Jupiter and Saturn. Five satellites, or moons, can also be seen circling round the planet.

Neptune and Pluto are so far away and so faint that they are invisible to the naked eye. A telescope shows Neptune as a bluish disc, with two moons circling round it. But even a powerful telescope will show Pluto only as a pinpoint of light.

Deep space *Voyager*

To find out more about the planets, astronomers have enlisted the help of space scientists. In August 1977, a space probe named *Voyager 2* blasted off the launch pad at Cape Canaveral in Florida. Its mission was to fly to the giant planets Jupiter and Saturn, then on to Uranus and Neptune. It was making use of a chance

△ *Voyager 2* pictures Neptune and its large moon Triton as it nears the planet.

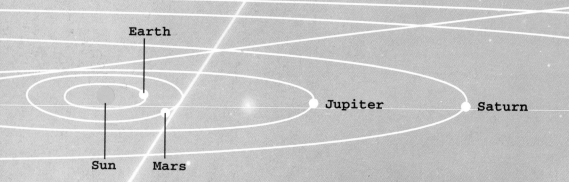

Earth

Jupiter   Saturn

Sun   Mars

Next stop, the stars

Neptune was *Voyager 2*'s last port of call. At the start of the new millennium, the probe was 9000 million kilometres away from the Earth. It is heading for interstellar space – the space between the stars. But it will not get near another star for tens of thousands of years.

line-up of the outer planets that would not happen again for 176 years. A sister craft, *Voyager 1*, set off a month later.

Both *Voyagers* visited Jupiter and Saturn in turn, and then *Voyager 1* began to head out of the Solar System. But *Voyager 2* sped on to Uranus, flying past the planet in January 1986. It made many new discoveries, showing sets of rings around the planet and a swarm of tiny new moons.

Three years later, *Voyager 2* homed in on Neptune, passing closest in August 1989. It sent back spectacular pictures of storms and spotted icy volcanoes erupting on Triton, Neptune's largest moon.

Eyes in orbit

Since the early 1990s, astronomers have continued to keep an eye on Uranus and Neptune using the Hubble Space Telescope (HST). The HST gets a much clearer view of the two planets than telescopes on Earth.

Pluto

◁ Uranus, Neptune and Pluto lie billions of kilometres from the Sun. Pluto lies 40 times farther away from the Sun than the Earth does. It is the only planet that space probes have not visited yet.

Uranus

Neptune

URANUS, THE NEW WORLD

△ **Uranus symbol**

On the night of March 13, 1781, William Herschel was stargazing. A musician by profession, he was a keen astronomer as well. William Herschel lived in Bath, in south-west England. That night he was looking at the stars in the constellation Gemini, the Twins. He noticed that one of the 'stars' appeared as a disc, not as a point, like all the other stars. It was, he wrote in his notebook, 'a curious either nebulous star or perhaps a comet'.

But the object was neither a curious star nor a comet. It was a new planet, the first planet to be discovered since ancient times. It was named Uranus. When the orbit of the new planet was calculated, it was found to lie more than 2800 million kilometres away from the Sun. It was twice as far away as Saturn.

△ **William Herschel**

A smaller giant

Uranus is the third largest planet, after Jupiter and Saturn. With a diameter of about 51 200 kilometres, it is less than half the size of Saturn. But it is still four times the size of the Earth.

Uranus is so far away that it takes 84 Earth-years to travel once around the Sun. From the Earth, we can watch it make its way slowly through the heavens against the background of stars. Like all the planets, it travels through the constellations of the zodiac.

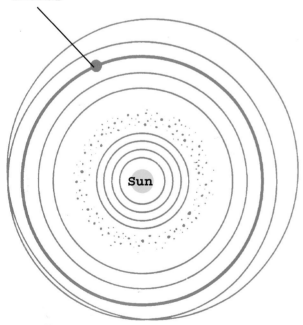

Uranus

Sun

△ **Uranus, the seventh planet out from the Sun, lies beyond Saturn.**

Uranus

△ Uranus is four
times bigger across
than our own planet.

Earth

All topsy-turvy

Every planet spins round like a top as it travels in its orbit around the Sun. Most planets spin round in a nearly upright position in relation to the direction they are travelling. Their axis – the imaginary line around which they spin – is nearly at right angles to the direction they travel.

Uranus is different. Its axis leans right over and lies nearly in the direction the planet travels. So, compared with the other planets, Uranus spins on its side. This means that, as it travels around the Sun, its poles face towards and away from the Sun in turn. Each pole spends 42 years in the sunlight, followed by 42 years in darkness.

△ In Greek mythology, Uranus was god of the heavens. He was killed by Saturn, who was his youngest son.

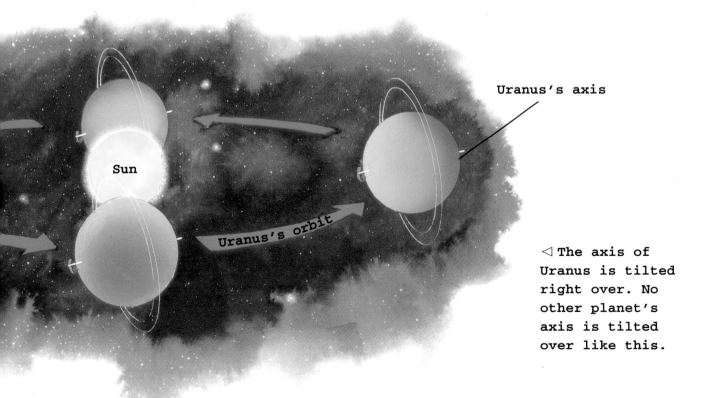

Sun

Uranus's orbit

Uranus's axis

◁ The axis of
Uranus is tilted
right over. No
other planet's
axis is tilted
over like this.

9

URANUS INSIDE AND OUT

Faint rings surround this greenish-blue gas giant.

Like Jupiter and Saturn, Uranus is made up mainly of gas and liquid. It has no solid surface. On top, it has a deep atmosphere of hydrogen and helium gases, along with a little methane. (On Earth, methane is the main gas in natural gas, which is piped into our homes for cooking and heating.)

It is the methane which gives Uranus's atmosphere its colour. The gas absorbs the red light from sunlight, making the atmosphere appear bluish-green.

Underneath the atmosphere is a deep, warm ocean of water, methane and ammonia. And at the centre of the planet is a core of solid rock.

Clouds and winds

When *Voyager 2* visited Uranus, the atmosphere looked featureless in ordinary light. There were no bands of clouds like those on Jupiter. The clouds of methane that must have been present were hidden by a thick haze.

Careful computer processing of the images made a few isolated clouds show up, together with faint bands that were presumably cloud belts. Winds forced the clouds along at speeds up to about 550 kilometres an hour. Since then, the Hubble Space Telescope has spotted strings of bright clouds on the planet.

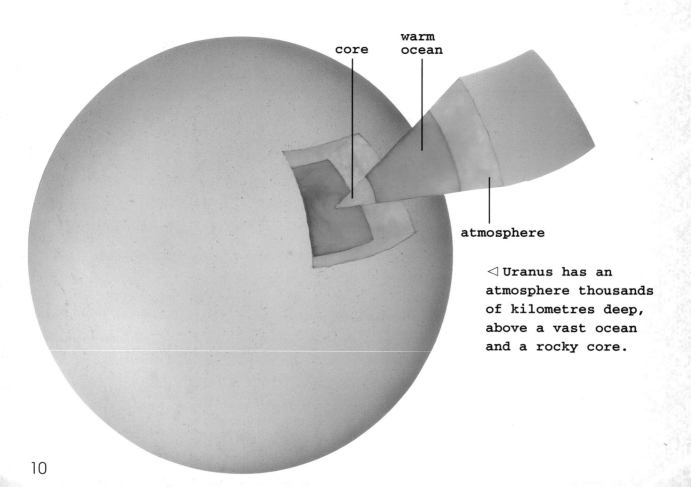

core

warm ocean

atmosphere

◁ Uranus has an atmosphere thousands of kilometres deep, above a vast ocean and a rocky core.

Rings around Uranus

In 1977, the year *Voyager 2* was launched, astronomers discovered that Uranus had a set of rings circling round it. This made Uranus the second planet known to have rings, after Saturn. But Uranus's rings are too faint to be seen from Earth. Nine rings in all were spotted when they passed in front of a star and one by one blocked its light.

 Voyager 2 discovered two more rings and pictured the whole ring system clearly. Compared with Saturn's broad, bright shining rings, the rings around Uranus are narrow and very dark – as black as coal. Most of the other rings are less than 10 kilometres wide. The outermost ring, the Epsilon, is the widest at up to 100 kilometres across.

▽ **In normal photographs, Uranus looks the same all over. But a false-colour image shows a hazy region (orange).**

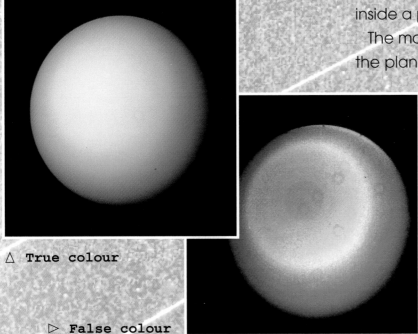

△ **True colour**

▷ **False colour**

Listen to the radio

Voyager 2 made another discovery when it flew past Uranus. It picked up radio signals from around the planet. To scientists on the Earth, this could only mean one thing – that Uranus has a magnetic field around it, just like the Earth and the giant planets. Radio signals are set up when electrically charged particles whizz about inside a planet's magnetic field.

 The magnetic field of a planet rotates as the planet itself rotates. By studying the radio signals the planet gives off, the time of rotation can be found. The radio signals from Uranus showed that the planet rotates once every 17 hours 14 minutes. This means that its 'day' is about 7 hours shorter than our own (24 hours).

Background image: Six of the ten rings around Uranus.

MANY MOONS

Uranus's icy moons are cracked and cratered.

From the Earth, only five of Uranus's moons can be seen in a telescope. Going out from the planet, they are Miranda, Ariel, Umbriel, Oberon and Titania. Except for Umbriel, the moons were named after characters in William Shakespeare's plays.

When the *Voyager 2* probe flew past Uranus in 1986, it discovered another ten tiny moons that were nearer the planet than Miranda. And in 1997 astronomers found two more moons very far out, making 17 in all. These new moons, too, have been given the names of characters in Shakespeare's plays.

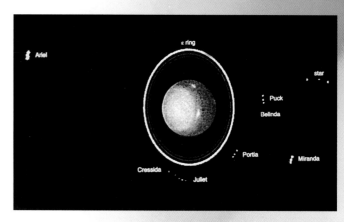

△ **This Hubble Space Telescope picture shows Uranus and seven of its moons.**

▽ **Deep valleys and steep cliffs scar the surface of Ariel.**

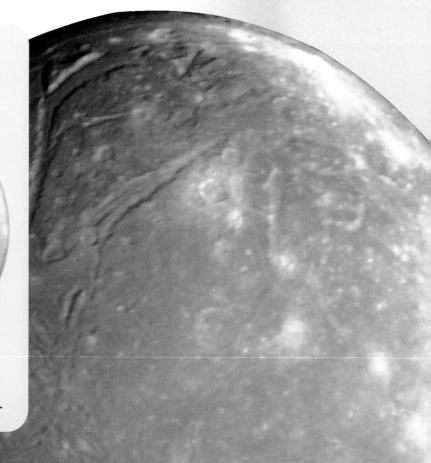

Little shepherds

The moons *Voyager 2* discovered really are tiny. Even the biggest, Puck, is only about 150 kilometres across. The two smallest, Cordelia and Ophelia, are only about 30 kilometres across. These two moons circle one each side of the outer ring of Uranus, the Epsilon. They are known as shepherd moons because of the way they seem to 'herd' the particles in the ring and keep them in place.

The old moons

The five large moons of Uranus seem to be made up of a mixture of rock and ice. Astronomers work this out from the density, or size and weight of the bodies. The moons may have a large core of rock with a layer of ice on top, or they may be made up of a jumbled mass of rock and ice. But they all seem to have an icy surface.

With a diameter of about 1580 kilometres, Titania is the biggest moon, slightly bigger than Oberon. Ariel and Umbriel are smaller, both about 1160 kilometres across. They are more than twice the size of Miranda.

Craters and cracks

Craters are found all over the surface of the moons. They have been made by meteorites smashing into the surface. Some of the craters are bright, which shows that they are recent. The meteorites have exposed fresh white ice in the otherwise darker, dirtier surface. The most craters are found on Oberon and Umbriel, so these two moons probably have old surfaces.

Valleys and canyons

Ariel and Titania have fewer craters and are cut by networks of valleys, or canyons, several kilometres deep. These canyons must have been formed when the crusts of the moons moved and cracked long ago. Volcanoes would then have erupted in the cracks, or faults. But they would not have given off hot liquid rock, but ice-cold water. The water would then have frozen, producing the smooth floors the canyons have today.

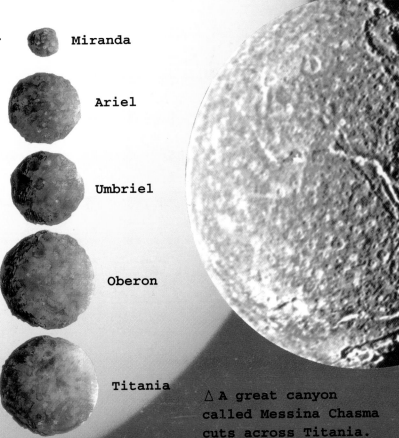

Miranda

Ariel

Umbriel

Oberon

Titania

▷ The five main moons of Uranus compared in size with the Earth's Moon, which measures 3476 kilometres across.

△ A great canyon called Messina Chasma cuts across Titania.

Earth's Moon

MIRANDA'S A MIRACLE!

Miranda's patchwork surface is the oddest in the Solar System.

By chance, when *Voyager 2* visited Uranus, the moon it passed closest to was Miranda. It flew within 3000 kilometres of this, the smallest of Uranus's moons that we can see from Earth. This was an amazing stroke of luck, because Miranda happens to be by far the most interesting of Uranus's moons and one of the strangest moons in the Solar System.

Miranda has all kinds of surface features jumbled up together, like the different patches in a patchwork quilt. Here is a patch of old cratered plains; there, a patch with young steep cliffs. Here is a peculiar V-, or chevron-shaped, patch; there, a strange circular grooved region that looks like a gigantic race-track.

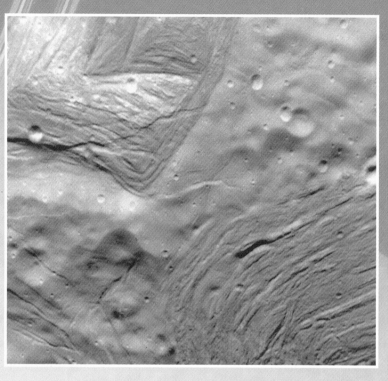

△ **Different kinds of features meet on Miranda's surface.**

Geologists baffled

Planetary geologists – the scientists who study the surface and make-up of other planets – had never seen anything like Miranda's surface before. Even now they are not sure how it came about.

One idea is that long ago Miranda suffered a catastrophe. It was hit by another huge object and was smashed to bits. In time, gravity pulled the bits together, and Miranda became a single moon once more. And the patches we now see are the tops of the separate bits that came together. The cratered plains are part of the surface of the old moon.

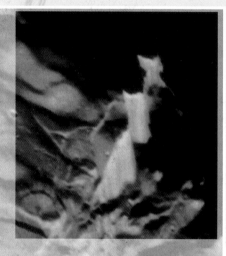

△ **Miranda's icy cliffs, called Verona Rupes, soar to a height of more than 20 kilometres.**

△ **If you were circling Miranda in a spacecraft, Uranus would loom large in the background.**

Miranda in close-up

We only know what one side of Miranda looks like, which is the side that *Voyager 2* pictured. About half of Miranda's surface is covered by old cratered plains. There are two large grooved regions of valleys and ridges, with the chevron-like area between them.

Astronomers have called these regions coronae, a word meaning crowns. They think that they formed as a result of movements of the crust, followed by infilling by material from icy volcanoes.

Parts of the moon are cut with deep canyons. In one place, the canyon walls soar to a height of 20 kilometres, or more than twice the height of the Earth's highest mountain, Mount Everest.

The darker groovy and chevron-shaped regions could be bits that came from the inside of the old moon.

A simpler explanation for Miranda's strange appearance is patchy activity on the surface. Movements in the crust rippled the surface. But they took place only over small areas, leaving the surrounding older cratered surface the same.

▷ **This is a mosaic of *Voyager* photos of Miranda. The rectangular feature in the middle is named Inverness Corona.**

NEPTUNE, THE BLUE WORLD

Neptune is a near twin of Uranus, both in size and make-up.

△ **Neptune symbol**

After William Herschel discovered Uranus in 1781, astronomers calculated what its orbit around the Sun should be. But they found from their observations that the planet did not accurately follow this orbit. Maybe, they thought, it was being pulled off course by the gravity of another planet farther out.

Several people tried to work out where this new planet might be found. Among them were two mathematicians, John Couch Adams in England and Urbain Leverrier in France. By September 1845,

▷ **The sea god Neptune**

Adams had worked out a position where he thought the new planet might be. Some eight months later, Leverrier picked on the same position, although he knew nothing of Adams's work.

At first, astronomers took little notice of both mathematicians. But on September 23, 1846, Leverrier wrote to Johann Galle at the Berlin Observatory, telling him where to look for the new planet. Galle spotted the planet the same night. It was later called Neptune. When its orbit was calculated, Neptune was found to lie more than 1600 million kilometres farther away from the Sun than Uranus, at a distance of nearly 4500 million kilometres.

Uranus's twin

With a diameter of about 49 500 kilometres, Neptune is only a fraction smaller than Uranus, making it the fourth largest planet in the Solar System. Being much farther away than Uranus, it takes nearly twice as long to circle once round the Sun – nearly 165 Earth-years.

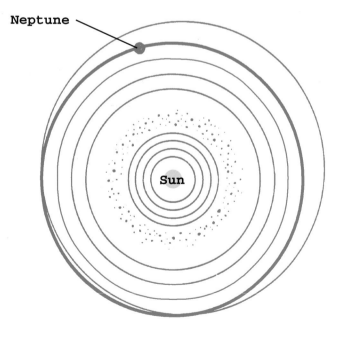

△ **Neptune is the eighth planet out from the Sun.**

Neptune

Earth

Viewed from the Earth, Neptune can be seen moving very slowly against the background of stars. At present it is travelling in the constellation of Capricornus. And it will remain in this constellation until the year 2010.

△ **The blue world Neptune is about four times bigger across than our own blue world.**

Heat from within

The temperature at the top of Neptune's atmosphere is a freezing -210°C. Strangely, this is the same temperature as at the top of Uranus's atmosphere. We would expect Neptune to be colder than Uranus because it is much farther away from the Sun.

So Neptune somehow produces heat inside it, just like Jupiter and Saturn. This internal heat probably explains why there is so much activity in Neptune's atmosphere.

In a spin

Neptune spins round in space like all the planets. But its spin axis is not tipped over like Uranus's. It is tilted only a little bit more than the Earth's. Neptune spins round once in 16 hours 7 minutes, which means that its 'day' is almost an hour shorter than Uranus's day.

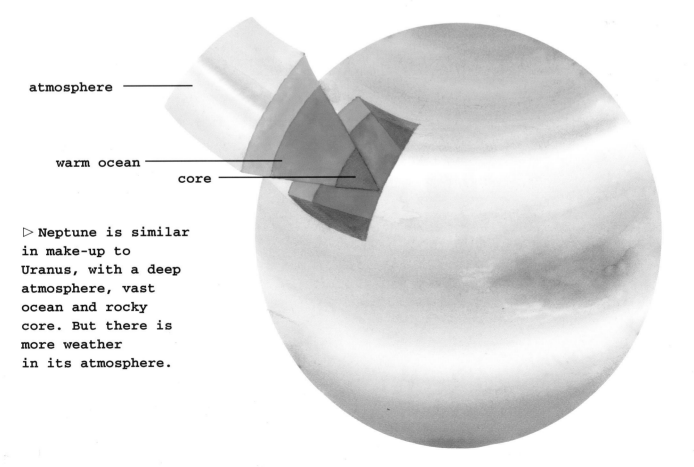

atmosphere ———————

warm ocean ———————
core ———————

▷ **Neptune is similar in make-up to Uranus, with a deep atmosphere, vast ocean and rocky core. But there is more weather in its atmosphere.**

Clouds form and storms rage in Neptune's windy atmosphere.

In make-up, Neptune seems to be almost identical to Uranus. It has an atmosphere made up mainly of hydrogen and helium, with some methane. As on Uranus, the methane makes the atmosphere appear bluish. Neptune's atmosphere is a deeper blue because there is not so much high-level haze.

The atmosphere seems to be about 5000 kilometres deep. Underneath there is a warm ocean consisting of a mixture of water, methane and ammonia. At the planet's centre is a core of rock, slightly bigger than the Earth.

Icy clouds

White wisps of cloud float high up in Neptune's deep blue atmosphere. They look rather like the high cirrus clouds in the Earth's atmosphere which we call mares' tails. Cirrus clouds are made up of tiny water ice crystals. And the wispy clouds on Neptune are probably made up of ice crystals too, but crystals of methane, not water.

Deeper down comes the main cloud layer, which seems to be made up of methane droplets. Overall the clouds are drawn out into bands parallel with Neptune's equator. This happens because the planet spins round so fast. Beneath the methane clouds, scientists expect there to be layers of other clouds – of ammonia, hydrogen sulphide, and even water. These would be similar to the cloud layers found on Jupiter and Saturn.

◁ This view of Neptune's southern hemisphere was produced from the pictures *Voyager 2* took in 1989. It shows the Great Dark Spot and nearby clouds.

▷ These pictures show changes in Neptune's atmosphere over three weeks.

◁ Bands of clouds in Neptune's atmosphere, highlighted by a low Sun.

Jet streams

Winds blow strongly in Neptune's atmosphere, moving in great streams parallel with the equator. Scientists call these streams zonal jets. The winds in the broad equatorial jet stream blow towards the west. Far to the north and south of the equator, the winds change direction and blow towards the east.

The strongest winds are found in the equatorial jet stream, where they can travel at speeds of more than 1500 kilometres an hour. This is about five times the speed of the strongest winds we have on the Earth (in tornadoes). In the whole Solar System, only Saturn has stronger winds.

Disappearing spots

When *Voyager 2* visited Neptune in 1989, it spied a huge dark oval in the southern atmosphere. The oval feature looked rather like the Great Red Spot on Jupiter, and so it was called the Great Dark Spot (GDS). Like Jupiter's spot, the GDS was a huge storm system.

The *Voyager* pictures also showed smaller dark spots, and a patch of white cloud nicknamed the Scooter, because it travelled so fast. When the Hubble Space Telescope began looking at Neptune in 1995, the spots had disappeared. But the atmosphere was much cloudier than it had been six years earlier.

RINGS AND MOONS

Neptune proves to be another ringed world.

When *Voyager 2* flew past Neptune, it discovered that this planet, too, has rings around it. This means that all the giant planets – Jupiter, Saturn, Uranus and Neptune – have rings. But only the rings of Saturn can be seen from the Earth.

Neptune's rings are very faint. They seem to be made up of tiny, very dark particles, rather like the smoke from a bonfire. In all there are five rings. The two brightest are named Adams and Leverrier, after the two mathematicians who worked out where Neptune could be found in the heavens (see page 16).

Adams and Leverrier are both about 50 kilometres wide. In Adams, the fine particles clump together in places to give brighter arcs, which have been spotted from Earth. The other three rings are much fainter but broader. They have also been named after people involved in the discovery of Neptune.

More moons

In their telescopes, astronomers on the Earth can see only two moons circling around Neptune. They are Triton and Nereid. Triton orbits quite close to the planet, but Nereid orbits very far out, at a distance of more than 5 million kilometres. Nereid is quite a small moon, with a diameter of only about 340 kilometres. Triton is much larger, with a diameter of about 2700 kilometres

Voyager 2 spotted six more moons when it flew past Neptune. Surprisingly, one proved to be larger than Nereid. Named Proteus, it measured about 400 kilometres across. This moon cannot be seen from the Earth because it orbits too close to Neptune and is always lost in the planet's glare.

Neptune's other five moons are much smaller. The largest, Larissa, is no more than about 200 kilometres across; the smallest, Naiad, is less than 60 kilometres wide.

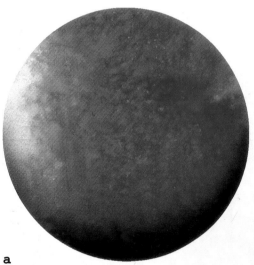

▷ **With a diameter of about 2710 kilometres, Triton is more than six times bigger than any of the other moons of Neptune.**

Triton

Proteus

Nereid

Larissa

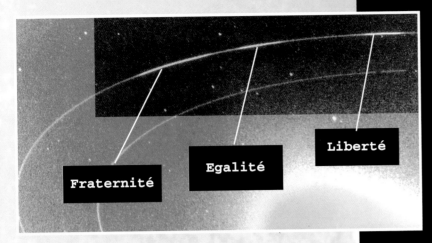

Liberté

Egalité

Fraternité

△ The bright arcs in Neptune's outer ring are named after a slogan of the French Revolution (meaning Brotherhood, Equality, Liberty).

Larissa circles Neptune outside the planet's rings, but the other moons circle within the ring system itself. Galatea circles just inside Adams, the outermost ring. It seems to be a lone shepherd moon, helping to keep the ring particles in place. Despina seems to be a shepherd for the other prominent ring, Leverrier.

△ Proteus, Neptune's second largest moon, is a shapeless lump, covered all over in craters.

▽ Neptune's rings have been named after prominent people involved in the planet's discovery.

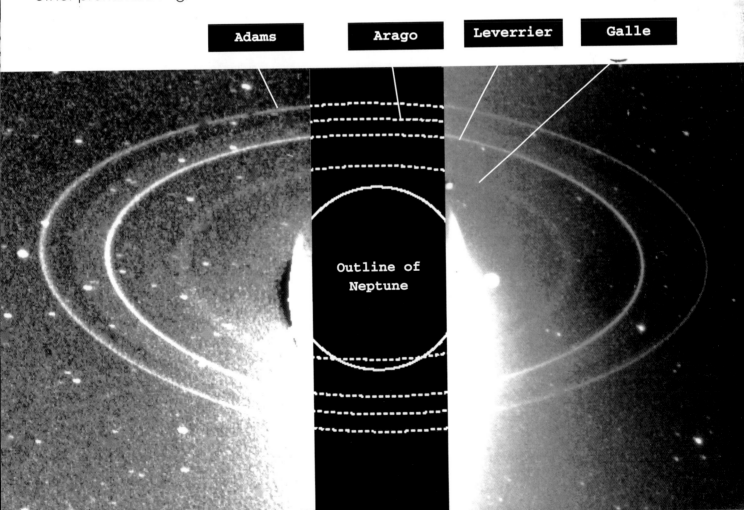

Adams

Arago

Leverrier

Galle

Outline of Neptune

COLD, COLD TRITON

Neptune's largest moon is full of surprises.

1. A distant view of Triton shows it has a pinkish surface.

2. Coming nearer, pinkish snow covers the moon's south polar regions.

An English astronomer named William Lassell discovered Triton in October 1846, only a few weeks after Neptune itself had been discovered. Even then, astronomers realized that the moon was special. All the moons of other planets they had discovered travelled in the same direction as the planets spun round. But they saw that Triton travelled in the opposite direction. Astronomers say that it has a retrograde (backwards) motion.

Triton is the only large moon in the Solar System to have a retrograde motion. And astronomers think that it might once have been an independent body circling in the depths of the Solar System. But long ago, it came too close to Neptune and was captured by the planet's gravity.

The deep freeze

Before *Voyager 2* flew past Triton, astronomers expected the moon would be a deep-frozen, icy world, pitted with craters, like some of the large moons of Saturn and Uranus. It would almost certainly be a 'dead' world, with no signs of geological activity going on because of the extreme cold. So they thought that Triton would probably not be a very interesting place.

They were right about it being deep-frozen and icy. The temperature on Triton is -235°C, which means that it is the coldest place we know in the whole Solar System.

It is even colder than Pluto, which is much farther out from the Sun. But the astronomers were wrong about Triton being dead and uninteresting. *Voyager 2* showed that Triton is very much alive and has an extraordinary surface.

Triton's surface

The first surprising thing about Triton is that it has very few craters. This indicates that the surface is relatively young, so processes must be going on that constantly renew the surface.

Astronomers think that icy volcanoes are at work. Liquid or slushy ice 'lava' flows onto the surface through cracks in the ground. This creates all kinds of features on Triton. There are large areas covered with ridges and circular dimples. This is called cantaloupe terrain, because it looks rather like the surface of a cantaloupe melon. There are also several smooth and flat plains, or planitia, and strange 'spots', or maculae, which are dark in the middle and brighter outside.

But the strangest features of all are erupting volcanoes blasting out dark 'smoke'. This is carried away by the wind in the faint atmosphere that covers the moon, leaving dark streaks on the pinkish ice over parts of the surface.

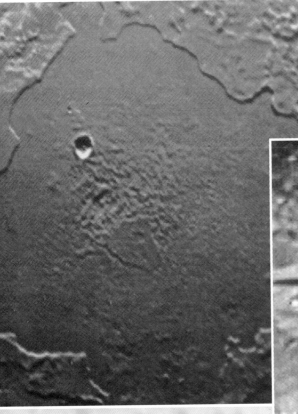

3. Closer in still, a wide, smooth basin, named Ruach Planitia, shows up.

4. These strange spots, called maculae, are as much as 200 kilometres across.

23

PLUTO, THE TINIEST WORLD

Icy Pluto is even smaller than the Moon.

△ **Pluto symbol**

After Neptune had been discovered, astronomers started looking for a ninth planet. An American astronomer named Percival Lowell worked out that the planet should be found travelling through the constellation Gemini, the Twins. But astronomers at his observatory in Flagstaff, Arizona, had not found it by the time Lowell died, in 1916, and they gave up the search.

The search for the ninth planet did not begin again until 1929. The job fell to an astronomer named Clyde Tombaugh.

After weeks of searching, he finally found it on February 18, 1930. It was almost exactly where Lowell had predicted. The dim, distant planet was named Pluto, after the god of the underworld and the dead.

△ **In Greek myths Pluto ruled the underworld.**

Small world

Pluto is a tiny world, only about two-thirds the size of our Moon. It is by far the smallest planet. It appears to be made up mainly of rock and water ice. There is also methane ice – methane gas that has frozen – on the surface, together with frozen nitrogen gas. These ices evaporate (turn to vapour) to give a slight atmosphere.

Little is known about what Pluto's surface is like. But the Hubble Space Telescope has sent back pictures that show both light and dark regions.

In slow motion

Pluto travels slowly in a long oval orbit, that takes it sometimes more than 7000 million kilometres from the Sun. It takes the planet nearly 250 years to circle the Sun once. As it travels in its orbit, Pluto also spins round slowly on its axis. It spins round once in 6 days 9 hours.

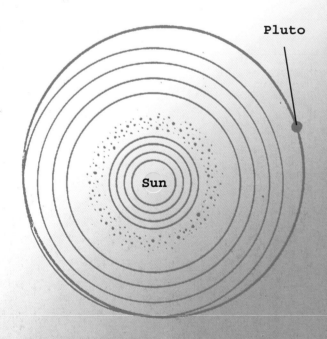

Pluto

Sun

△ **For most of the time, Pluto is the farthest planet from the Sun.**

▽ Pluto is twice the size of Charon.

Pluto   **Charon**

Big moon

Although Pluto is smaller than some moons in the Solar System, it has a moon itself. This moon, named Charon, is half the size of Pluto and orbits only about 18 000 kilometres high. It circles round Pluto in 6 days 9 hours, or in exactly the same time Pluto takes to spin round once. This means that Charon always stays over the same spot on Pluto.

Charon was discovered by the American astronomer James Christy in 1978. Nine years later, astronomers were lucky enough to see Charon and Pluto eclipse, or pass in front of, each other. This will not happen again for over 100 years. Charon seems to be covered mainly with water ice. Its surface is also darker than that of Pluto.

△ Pluto and Charon, photographed by the Hubble Space Telescope.

▽ Pluto probably has quite a large core of rock, with a thick layer of water ice on top. Other ices, such as methane, cover its surface.

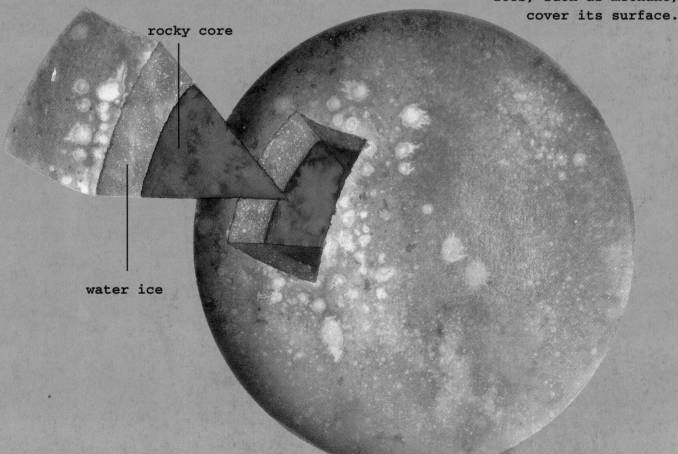

rocky core

water ice

25

PLANETS BEYOND

Astronomers are looking for new planets beyond Pluto and around other stars.

The search for new planets did not end with the discovery of Pluto in 1930. Astronomers soon realized that Pluto was too small a body to have any effect on the orbits of Uranus and Neptune. So ever since, they have been looking for another body – planet number ten, or Planet X.

In 1999, some astronomers announced that they had found evidence of a much more distant planet. This planet, probably at least as big as Jupiter, circles the Sun at a distance of 3 to 5 million million kilometres, hundreds of times farther out than Pluto.

The astronomers worked out that this planet must exist by studying the orbits of comets.

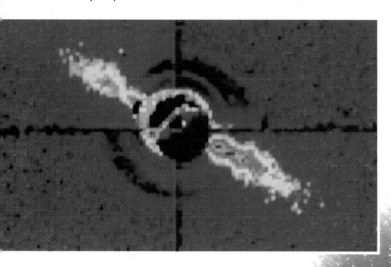

△ **There is a dusty disc around the star Beta Pictoris.**

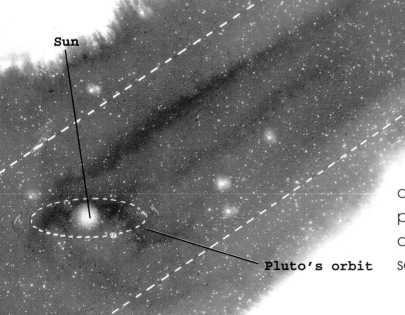

Sun

Pluto's orbit

Comets are found in a huge cloud millions of millions of kilometres away. They start to travel in towards the Sun when something disturbs them. The astronomers think that it is the distant planet that does this. But even if the planet does exist, there seems little chance of seeing it because it is so far away.

possible orbit
of distant planet

cloud of comets

△ **A very distant
planet may circle
in or near the Oort
cloud of comets.**

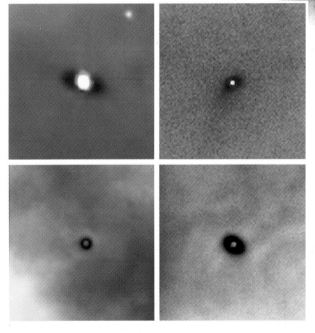

△ **The Hubble Space Telescope
has spied dusty discs
(proplyds) around four stars
in the Orion nebula.**

Planets of other stars

The planets in our Solar System, which circle around the Sun, are not the only planets that exist. There are millions upon millions of stars like the Sun far away in space. And most of these will probably have planets circling around them. Planets are born in the discs of gas and dust that surround most stars when they form. The Hubble Space Telescope has photographed many such discs, which are known as proplyds, or protoplanetary discs.

Astronomers already know of more than ten stars with planets circling round them. They call them extrasolar planets, meaning planets outside our Solar System. These planets are too far away to be seen, but astronomers know they are there because they cause the stars to wobble slightly.

Uranus data

Diameter at equator:	51 200 km
Volume:	67 times Earth's volume
Mass:	15 times Earth's mass
Density:	1.3 times density of water
Gravity at surface:	1.2 times Earth's gravity
Distance from Sun	
average:	2 870 000 000 km
farthest:	3 004 000 000 km
closest:	2 735 000 000 km
Spins on axis in:	17 hours 14 minutes
Circles Sun in:	84 years
Speed in orbit:	24 500 km an hour
Temperature:	-210°C
Moons:	17

Pluto data

Diameter at equator:	2284 km
Volume:	0.002 times Earth's volume
Mass:	0.0005 times Earth's mass
Density:	2 times density of water
Gravity at surface:	0.04 times Earth's gravity
Distance from Sun	
average:	5 900 000 000 km
farthest:	7 375 000 000 km
closest:	4 425 000 000 km
Spins on axis in:	6 days 9 hours
Circles Sun in:	248 years
Speed in orbit:	17 000 km an hour
Temperature:	-220°C
Moons:	1

Neptune data

Diameter at equator:	49 500 km
Volume:	57 times Earth's volume
Mass:	17 times Earth's mass
Density:	1.8 times density of water
Gravity at surface:	1.2 times Earth's gravity
Distance from Sun	
average:	4 497 000 000 km
farthest:	4 537 000 000 km
closest:	4 456 000 000 km
Spins on axis in:	16 hours 7 minutes
Circles Sun in:	164.8 years
Speed in orbit:	19 500 km an hour
Temperature:	-210°C
Moons:	8

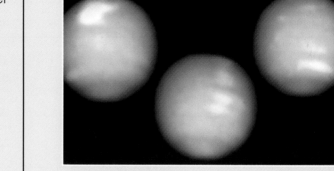

URANUS NOTES

AT FIRST SIGHT

Before William Herschel discovered Uranus in March 1781, the planet had been observed several times as an ordinary 'fixed star', first by the English astronomer John Flamsteed in 1690. Flamsteed had been appointed England's first Astronomer Royal by King Charles II in 1675.

SWEET DREAMS

The discoverer of Uranus, William Herschel, was also first to spot two of its moons, Titania and Oberon, in January 1787. Herschel's son John suggested that the moons of Uranus should be called after characters in Shakespeare plays.

THE WINKING STARS

On March 10, 1977, some American astronomers were flying over the Indian Ocean in the Kuiper Airborne Observatory, a plane equipped with a powerful telescope. They were planning to watch an occultation (covering up) of a star by Uranus so that they could work out the planet's diameter accurately. Just before the occultation was due, the star 'winked' five times. Something near Uranus was passing in front of the star. It also 'winked' five times afterwards. Similar 'winks' were also recorded at an observatory on the ground. Astronomers decided that the 'winks' were caused by a set of rings around the planet. In 1986, the *Voyager 2* probe took photographs of these rings.

NEPTUNE NOTES

BAD LUCK

By September 1845, the English mathematician John Couch Adams had worked out where he thought an eighth planet would be, far beyond Uranus. He sent his calculations to the English astronomer James Challis. Challis did not bother to follow the matter up. If he had looked for the new planet where Adams had suggested, he would have spotted it a year before its actual discoverer, Johann Galle. Challis eventually started a search in July 1846, but with no great urgency or careful checking. When he heard of Galle's discovery, he rechecked his July observations and found that he had spotted the new planet without realizing it!

WHAT'S IN A NAME?

The discoverer of the new eighth planet, Johann Galle, suggested the name 'Janus' for it. Jean Leverrier, who had calculated the planet's position at much the same time as Adams, suggested 'Neptune', but changed his mind and suggested 'Leverrier'. But 'Neptune' was more in keeping with the names of other planets and came to be accepted.

PLUTO NOTES

TWICE UNLUCKY

In 1919, the American astronomer Milton Humason (who first started work as a donkey-driver) began a search for a ninth planet. Like others, he looked at photographic plates to see if any objects appeared where they should not among the known stars. After Pluto had been discovered 11 years later, the plates Humason had used were checked. He should have found Pluto on one in 1919, but the image of the planet happened to fall on a flaw in the plate and so was not spotted. And on another plate, Pluto happened to be passing in front of a star!

NOW YOU SEE IT...

Charon circles round Pluto in exactly the same time that Pluto spins round on its axis. This means that if you lived on one side of the planet, you would always see Charon in the same spot in the sky. But if you lived on the other side, you would never see the moon at all.

TIME LINE

1690

England's first Astronomer Royal, John Flamsteed, records the then unknown planet Uranus as a star in the constellation Taurus.

1750

The French astronomer Pierre Le Monnier spots Uranus for the first of eight times without realizing it is not a star.

1781

English musician turned astronomer William Herschel discovers Uranus on March 13. The German astronomer Johann Bode suggests the name of the new planet.

1787

William Herschel discovers two of Uranus's moons in January. They are Oberon and Titania, names suggested by William's son John.

1845

The English mathematician John Couch Adams calculates the position of a possible eighth planet. So does Urbain Leverrier in France.

1846

The English astronomer James Challis starts looking for the eighth planet in July. He misses discovering it because he doesn't check his sightings against star charts carefully enough. The German astronomer Johann Galle spots the new planet exactly where Leverrier has predicted, on September 23. It becomes known as Neptune. The English astronomer William Lassell discovers Neptune's first satellite, Triton, in December.

1905

The US astronomer Percival Lowell begins a two-year search for a ninth planet at his Flagstaff Observatory.

1930

The US astronomer Clyde Tombaugh discovers the ninth planet on February 18, finding it on photographic plates exposed on January 21, 23 and 29. The planet is called Pluto.

1949

The US astronomer Gerard Kuiper discovers a second moon of Neptune, Nereid.

1978

US astronomers find rings around Uranus on March 10, flying in the Kuiper Airborne Observatory. The US astronomer James Christy discovers that Pluto has a moon, Charon.

1979

Neptune becomes the most distant planet, as Pluto moves inside Neptune's orbit.

1986

Voyager 2 flies past Uranus on January 24, approaching to within 80 000 km. It pictures the rings clearly and discovers new moons.

1989

Voyager 2 flies past Neptune on August 24, skimming within 5000 km of the cloud tops. It discovers rings and new moons.

1999

Pluto moves outside Neptune's orbit and becomes the most distant planet. Astronomers suggest that there is a planet far beyond Pluto that makes comets travel in towards the Sun.

GLOSSARY

atmosphere
The layer of gases around the Earth or another planet.

axis
An imaginary line through the middle of a spinning body, around which the body spins.

comet
A lump of ice and dust that starts to shine when it gets near the Sun.

constellation
A recognizable group of bright stars in the sky.

core
The centre of a body, such as a planet or a moon.

crater
A circular pit in the surface of a planet or moon. Most craters are made by meteorites, but some are the mouths of ancient volcanoes.

crust
The hard outer layer of a planet or a moon.

extrasolar planet
One that circles around another star.

fault
A crack in the surface of a planet or a moon, caused by movements in the body's crust.

gas giant
One of the giant planets – Jupiter, Saturn, Uranus and Neptune – which are made up mainly of gas and liquid gas.

gravity
The force of attraction that every body has because of its mass.

heavens
Another word for the starry night sky.

ice
Frozen water or frozen gas.

interplanetary
Between the planets.

magnetic field
The region around a planet in which its magnetism acts.

meteorite
A piece of rock or metal from space that hits a planet or a moon.

moon
The common name for a satellite of a planet.

naked eye
'With the naked eye' means using your eyes alone.

orbit
The path in space one body follows when it circles around another, such as Neptune's orbit around the Sun.

planet
One of the nine bodies that circle in space around the Sun.

probe
A spacecraft sent to explore other heavenly bodies, such as planets and their moons.

proplyd
Short for protoplanetary disc; a disc of matter found around young stars, which might one day collect together to form planets.

retrograde
Movement in the opposite direction from usual.

rings
Rings of particles that circle around the giant planets.

satellite
A small body that orbits around a larger one; a moon. Also the usual term for an artificial satellite, a spacecraft that orbits around the Earth.

seasons
Regular changes in temperature and weather caused by the tilt of a planet's axis in space.

shepherd moon
One that appears to keep the particles in a planet's ring in place.

solar
To do with the Sun.

Solar System
The Sun and the family of bodies that circle around it, including planets, comets and asteroids.

solar wind
A stream of particles that the Sun gives off.

Universe
Space and everything in it – galaxies, stars, planets and energy.

zodiac
A band in the heavens through which all the planets seem to travel. The band passes through 12 constellations – the constellations of the zodiac.

INDEX